José Mariano da Silva Neto

Catalytic hydrolysis of sisal fibre using atapulgite clay

José Mariano da Silva Neto

Catalytic hydrolysis of sisal fibre using atapulgite clay

A case study

ScienciaScripts

Imprint

Any brand names and product names mentioned in this book are subject to trademark, brand or patent protection and are trademarks or registered trademarks of their respective holders. The use of brand names, product names, common names, trade names, product descriptions etc. even without a particular marking in this work is in no way to be construed to mean that such names may be regarded as unrestricted in respect of trademark and brand protection legislation and could thus be used by anyone.

Cover image: www.ingimage.com

This book is a translation from the original published under ISBN 978-613-9-64166-6.

Publisher:
Sciencia Scripts
is a trademark of
Dodo Books Indian Ocean Ltd. and OmniScriptum S.R.L publishing group

120 High Road, East Finchley, London, N2 9ED, United Kingdom
Str. Armeneasca 28/1, office 1, Chisinau MD-2012, Republic of Moldova, Europe
Printed at: see last page
ISBN: 978-620-7-72744-5

I dedicate this work to my mother, Marineis Soares, who through her integrity and perseverance was able to guide me along the best paths, never forgetting ethical and moral principles.

My energy is the challenge, my motivation is the impossible, and that's why I need to be, by force and at will, unshakeable. (Augusto Branco)

SUMMARY

CHAPTER 1   5

CHAPTER 2   8

CHAPTER 3   22

CHAPTER 4   27

CHAPTER 5   35

SUMMARY

The production of chemicals and fuels from lignocellulosic materials has received particular interest from researchers due to the negative impact of fossil fuels on the environment, as well as the growing concern about geopolitical oil issues around the world. Current research shows the production of second-generation ethanol from lignocellulosic materials as an alternative technology. Sisal fibre is a lignocellulosic material composed basically of cellulose (58%), hemicellulose (15%) and lignin (11%), making it an alternative for second-generation ethanol production in north-eastern Brazil, as it is a crop with a considerable carbohydrate content that can be degraded into fermentable sugars. In this context, the aim of this work was to study the process of catalytic hydrolysis of sisal fibre using atapulgite clay as a catalyst for the production of these sugars. The atapulgite clay was characterised using X-ray diffraction (XRD) and physical adsorption of N2 by BET, and the sisal fibre was characterised in terms of moisture content, ash, extractives, lignin, holocellulose, alpha-cellulose and hemicellulose. An acid pre-treatment was carried out followed by a base treatment and the residue from this pre-treatment was hydrolysed using a 2^3 experimental design where the variables were: temperature, substrate mass and reaction time. The results of the XRD analyses showed that the clay had good crystallinity by the intensity of the reflection $d_{(231)}$ and the physical adsorption of N2 by BET showed that after acid activation with hydrochloric acid the clay had greater surface area and pore volume, 192.7 m /g and 0.3084 cm /g, respectively. The response surface generated by the experimental design showed that by setting the substrate mass at 1.5 g, a temperature of 200°C and a reaction time of 6 hours, the maximum amount of reducing sugars was obtained in the hydrolysed liquor, approximately 9,684.0 mg/L.

Keywords: lignocellulosic material, pre-treatment, reducing sugars.

CHAPTER 1

INTRODUCTION

The production of fuels from lignocellulosic materials has received significant interest in the academic world due to the negative impact of fossil fuels on the environment and also due to the growing geopolitical concern regarding the distribution of oil, the most widely used energy source (SÁ; CAMMAROTA; FERREIRA-LEITÃO, 2014).

According to a report released by the Energy Research Company (EPE) in 2014, total ethanol production from sugar cane stood at 29 billion litres, making it the renewable fuel with the highest production in Brazil. However, experts predict that this material will not be able to meet the demand that ethanol may require in the near future. With this in mind, researchers have turned their attention to ethanol production using lignocellulosic materials for two fundamental reasons: they won't compete with food crops and they are less expensive than conventional materials.

In view of this, research has been carried out to obtain sugars that can be biochemically transformed into ethanol. Leão (2014), for example, studied the potential for producing ethanol from sisal fibre using acid and enzymatic routes to produce sugars. Li *et* al (2014) studied the production of these sugars through ultrasonic enhancement using HCl-FeCh as a catalyst. Silva et al (2012) studied the hydrolysis of cellulose using mesostructured NiO-MCM-41 and MoO_3 -MCM-41 catalysts.

The process of converting lignocellulosic material into ethanol involves four stages: pre-treatment, hydrolysis, fermentation and distillation. The most commonly used processes for obtaining fermentable sugars from lignocellulosic biomass is through acid or enzymatic hydrolysis. However, various studies published in academic circles have shown that acid hydrolysis has generated low rates of sugars and is not a feasible process for ethanol production, and the problem with enzymatic hydrolysis lies in the high costs of enzymes that are capable of degrading the cellulose molecule into fermentable sugars such as glucose, for example.

Although there are several studies in the literature on the degradation of carbohydrates into sugars using the two routes mentioned above, little has been done on the use of clays as catalysts in the hydrolysis of lignocellulosic materials.

Sisal appears to be an ideal raw material for ethanol production due to its relatively low biomass cost and the fact that it is not used as a food source and, at the same time, has a high cellulose content. As this fibre is rich in cellulose (67-78%), hemicellulose (10-4.2%) and lignin (8-11%), according to Leão (2014), a study of this material, which is widely cultivated in the state of Paraíba and other states in the Brazilian Northeast, would add some value to it, making it a very promising alternative for the production of a clean and sustainable energy source.

In this context, this work aims to study the process of hydrolysis of sisal fibre using atapulgite clay as a catalyst for the degradation of carbohydrates into fermentable sugars.

1.1 OBJECTIVES

1.1.1 General objective

To evaluate the potential of atapulgite clay in the hydrolysis of sisal fibre (Agave *sisalana)* for the production of fermentable sugars.

1.1.2 Specific objectives

- Through the characterisation of *fresh* and pre-treated sisal fibres, the contents of moisture, ash, extractives, holocellulose, alpha-cellulose, hemicellulose and lignin were quantified;
- Carry out an acid treatment on the atapulgite clay;
- X-ray diffraction was used to identify crystallinity and physical adsorption of N_2 by BET to quantify the surface area and pore volume of the natural and treated clays;

- Carry out an acid treatment followed by an alkaline treatment on the sisal fibre;

- To evaluate the catalytic hydrolysis of sisal fibre in the presence of atapulgite clay to obtain fermentable sugars.

CHAPTER 2

LITERATURE REVIEW

2.1 RENEWABLE ENERGY IN BRAZIL

Renewable energy is, by definition, sustainable and clean, and it also offers the opportunity to tackle the growing decline in the use of fossil resources and the impacts associated with them (CHEN *et* al, 2010). Globally, Brazil is in a favourable position when it comes to renewable energy sources. In 2013, more than 40 per cent of all primary energy produced in Brazil came from renewable energy sources (EPE, 2014), a relatively high figure compared to the world average of approximately 13 per cent (IEA, 2013).

Most of the renewable sources used in the country come from sugar cane products (16.1 per cent), hydroelectric power (15.5 per cent) and other biomass (8.3 per cent). Wind, solar and other renewable sources are still not very widespread, accounting for less than 5% of the total primary energy produced in Brazil (EPE, 2014).

In Brazil, a 5.3% increase in annual energy demand is expected over the next 10 years, reaching 372 million TEP (tonnes of oil equivalent) by 2020 (TOLMASQUIM, 2012). Even with the great diversification of the Brazilian energy matrix, oil is still the country's main energy source, with a share of 37.8 per cent (FERREIRA, 2007).

For this reason, research has been carried out to reduce the use of this fossil fuel, as it presents delicate issues in world geopolitics and also due to the great environmental concern and the emissions of atmospheric pollutants through the use of this type of fuel.

According to The Economist (2009), the Brazilian economy is expected to expand from the ninth to the fifth largest in the world by 2025, and knowing that the country has great potential in terms of natural resources and the expansion of an agro-industry, the production of biofuels, from an economic, environmental and social point of view, is seen as an attractive alternative for the production of renewable energy in Brazil.

In Brazil, for the production of ethanol, the main fuel produced in the country, sugar cane is currently the main raw material used, with 1 tonne of cane being enough to produce 85 litres of alcohol (RODRIGUES, 2010). Today, ethanol is the most efficient in the world, corresponding to 18% of the country's energy matrix, with annual production of 26 billion litres. According to Márcio Zimmermann, the former Minister of Mines and Energy, estimates suggest that by 2019 ethanol production will reach 64 billion litres per year, more than double current production.

In this scenario, Brazil has the potential to expand its ethanol production in the future (CHUN *et* al, 2011). This potential includes expanding the cultivation area, improving agriculture and introducing new industrial processing routes (LEITE, 2009). These new industrial processing routes can include the improvement of first generation ethanol (fermentation of sugars) as well as second generation ethanol (conversion of lignocellulosic biomass).

Second-generation industrial processing could greatly improve year-round ethanol production, as the supply of lignocellulosic crops is less seasonal than sugar cane. Despite ongoing research into second-generation technology, the economic performance of future technological developments remains uncertain (CHOVAU, 2013). Alongside sugar cane, other biomass feedstocks have been proposed for ethanol production due to their yield potential, composition or tolerance to climate and/or soil characteristics in other growing areas (BNDS, 2008).

2.2   SISAL: A BRIEF OVERVIEW OF BRAZIL

Sisal (Figure 1) is the main agro-industrial product of the Brazilian semi-arid region, which is currently plagued by historic drought. The crop is one of the few that can be grown in the region and shows visible signs of diminishing supply and intense signs of growing demand (NAVES, 2012).

Figure 1 - Sisal plantation

Source: AgroIntemational Ltd

Available at http://agrointemationalltd.com/ Accessed June 2015

Brazil is the world's largest producer and exporter of the fibre, according to data from the *Food and Agriculture Organization* (FAO), with around 50% of the world's production in 2009. In 2011, Brazil produced 111,000 tonnes. Bahia produced 95.8 per cent of this total. The states of Paraíba (3.5%), Ceará (0.4%) and Rio Grande do Norte (0.3%) are also producers. Historically, around 80% of production is commercialised to around 100 countries (AQUINO, 2012).

Sisal is predominantly grown in the north-east of Brazil, in semi-arid regions where there are few natural resources to grow other crops all year round. As a result, for many families in this region, sisal becomes a way of generating income during periods when there are no other crops to guarantee the minimum necessary for subsistence. It is against this backdrop that the sisal production process begins, where it is cultivated by small farmers who develop family agriculture on small rural properties (CUNHA, 2010).

Sisal leaves produce a highly resistant fibre, Figure 2, which is used to make handicrafts, brooms, bags, hats, string, ropes, mats and carpets, as well as to make cellulose for the production of Kraft paper (high resistance) and other types of fine paper (for cigarettes, filters, dielectric paper, sanitary napkins, nappies, etc.).

Figure 2 - Sisal fibres
Source: AgroIntemational Ltd Available at http://agrointemationalltd.com/ Accessed June 2015

In addition to these applications, sisal fibre can also be used in the automotive, furniture, household appliance and geotextile industries (for use in slope protection, agriculture and road surfacing), as a blend with polypropylene, in

replacing glass fibre (in the composition of plastic objects) and in construction. (CAMPBELL, 2004).

2.3   ARGYLS

Clays originate from alterations in igneous, metamorphic and sedimentary rocks; these alterations are caused by the chemical action of water, sulphur gases and weathering (GEREMIAS, 2003). They are made up of clay minerals, which are hydrated silicates of Al, Fe and Mg, whose basic structure consists of an octahedron in combination with tetrahedral silica (SiO_4) arranged to form the surface charge.

Clays are considered to be heterogeneous materials and their composition will depend on the location of extraction and the geological formation of the site. It is necessary to identify the type of clay to be used and its properties, as the characteristics of the raw material extracted are related to the properties of the final product (DUTRA *et* al.,2006, IBORRA *et* al.,2006).

Among the many uses of clay that are currently known are the manufacture of ceramic tiles, crockery, household utensils and decorative items, and they can also be used in aesthetic and health treatments. Recently, researchers have attributed yet another use to clays, which is their use as a catalyst in hydrolysis, using them as solid catalysts in aqueous media, since this way there are fewer problems associated with the environment and no chemicals are used in this process.

2.3.1 Atapulgite clay

Atapulgite or palygorschite is a hydrated magnesium-aluminium clay mineral with microfibrous morphology, low surface charges and high surface area (Neaman & Singer, 2004). Its octahedral layers extend along the entire length of the fibre and are continuous in only one dimension. Furthermore, its fibrous morphology is the result of the inversion that occurs every four silica tetrahedrons (BRIGATTI et al, 2006).

The formula for the unit cell of atapulgite is Mg_{10} Si O_{1640} $(OH)_4$.(OH $)_{28}$.$8H_2$ O, where OH2 indicates structural water molecules and H_2 O means the water present in the fibrous microchannels (COELHO et d, 2007). A representation of this structure is shown in Figure 3.

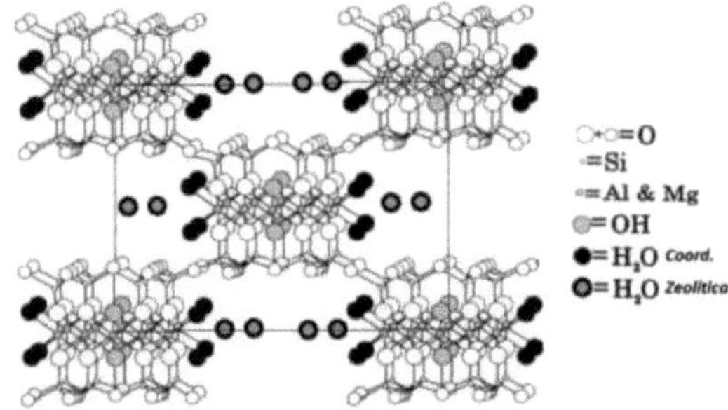

Figure 3-Details of the structure of atapulgite
Source: Xavier *et* al., 2012.

The main physicochemical properties of atapulgite are: high specific surface area estimated at around 125 to 210 m /g, high adsorption capacity, bleaching power, chemical inertness, does not swell in water, low surface charge due to small isomorphic substitution, cation exchange capacity of between 20-40 meq/100g (LUZ; BALTAR; OLIVERIA, 2003; MURRAY, 2000).

Due to these physicochemical properties, these clays can be applied for different purposes. Soares (2013) reports on the efficiency of atapulgite clay as a pharmaceutical excipient in solid forms, Sátiro (2013) evaluated the efficiency of this same clay as a catalyst for biodiesel production, Santos (2013) researched the potential of atapulgite from Guadalupe-PI as a lead ion adsorbent, due to the high capacity this ion offers the environment and Silva (2010) focussed on biodegradable polyesters - PHBV - as substitutes for oil-derived polymers by adding a small amount of atapulgite to them.

2.4   LIGNOCELLULOSIC FIBRES

Lignocellulosic raw materials are the most abundant renewable sources found in nature, mainly comprising agro-industrial materials, urban waste and the wood of angiosperms and gymnosperms (CASTRO et al., 2010).

Lignocellulose is a rigid complex made up of different molecules that constitutes almost the entire structural portion of plants, i.e. the largest percentage of plant biomass. It is a low-cost substrate that is considered to be final industrial waste and

The compositions of these materials vary and consist mainly of cellulose (35-50%) followed by hemicellulose (20-30%) and lignin (10-25%) and can be transformed into energy and chemicals (KUMAR *et* al., 2009).

Cellulose with the formula $(C\,H\,O_{6105})n$ is a C_6 polysaccharide made up of a long chain of glucose molecules. Hemicellulose with the formula $(C\,H\,O_{584})$ and $(C\,H\,O_{6105})$ is a relatively amorphous component that is more easily broken down chemically by heat than cellulose and is made up of a mixture of Cg polysaccharides (galactose and mannose) and C_5 polysaccharides (xylose and arabinose). Lignin is essentially the cement that provides the structural rigidity of plants and trees, formed by a three-dimensional polymeric network of methoxyl units, arylpropanes and hydroxyphenols. The empirical formula of this complex polymer is $C\,H\,O_{9102}\,(OCH)_{3n}$ in which n is the ratio of $CH_3\,O$ to C_9 groups: n = 1.4; 0.94 and 1.18 for hardwoods, softwoods and grasses, respectively. What gives this polymeric network its rigidity are the cross-links. Lignin is the largest non-carbohydrate constituent, accounting for between 15 and 25 per cent of the plant (RODRIGUES, 2011).

The basic composition of lignocellulosic biomass depends on the plant of origin, as shown in Table 1, and, in the case of agroforestry biomass, on the region, age and collection period of the material. According to Lima & Rodrigues (2007), the choice of biomass-producing species is a fundamental step in the production of cellulosic ethanol. Its productivity, cellulose and hemicellulose productivity and the productivity and type of lignin present must be taken into account.

Table 1 - Chemical composition of lignocellulosic biomass

Lignocellulosic Biomass	% Cellulose	% Hemicellulose	% Lignin
Sugarcane straw	40-44	30-32	22-25
Sugarcane bagasse	32-48	19-24	23-32
Wheat straw	30	50	15
Sisal	73,1	14,2	11
Maize fodder	38-40	28	7-21
Corn on the cob	45	35	15
Barley straw	31-45	27-38	14-19
Coconut fibre	36-43	0,15-0,25	41-45

Source: Gomez et al., (2010)

2.4.1 Cellulose

Cellulose, the main component of the cell wall of plant fibres, is a linear polymer containing up to 15,000 β-D-glucose units linked by β-1.4 carbon-carbon glycosidic bonds and by intramolecular and intermolecular hydrogen bonds, as shown in Figure 4. According to Yang *et* al (2010) it is the most abundant organic material on Earth, consisting of around 50% of all biomass and an annual production of approximately 100 billion tonnes.

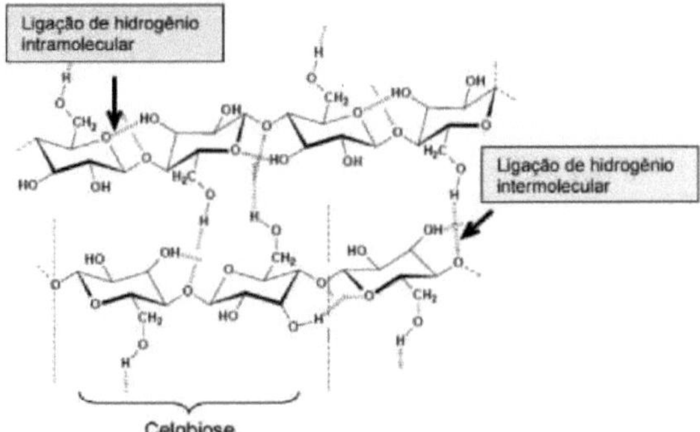

Figure 4 - Chemical structure of cellulose
Source: Morais, 2005

Intermolecular bonds are responsible for rigidity, and intramolecular bonds are responsible for the formation of fibrils, highly ordered structures that associate to form cellulose fibres. These characteristics, together with the lignin envelope, give the cellulose macromolecule great resistance to hydrolysis, which represents a major challenge for the use of lignocellulosic materials in biotechnological applications, such as the production of second-generation ethanol (ARANTES & SADDLER, 2010).

2.4.2 Haemicellulose

Hemicelluloses are another type of essential component present in the plant cell wall and are closely related to cellulose. They are sugar polymers that normally make up 20 - 40 per cent by weight of plant biomass, the composition of which can appear condensed in varying proportions. Unlike cellulose, hemicelluloses are not chemically homogeneous, have a low molar mass and do not contain crystalline regions, making them more susceptible to hydrolysis under milder conditions and varying in their structure and composition depending on the natural source (HUBER *et* al. 2006).Table 2 summarises the main differences between cellulose and hemicellulose.

Table 2 - Differences between cellulose and hemicellulose

Cellulose	Haemicellulose
Glucose units joined together	Units of different pentoses and hexoses linked together

High degree of polymerisation (1000 to 15000 glucose units)	Low degree of polymerisation (60 to 300 sugar units)
Forms fibrous arrangement It has amorphous and crystalline regions	Does not form a fibrous arrangement Only amorphous regions
It is slowly attacked by hot dilute inorganic acid	It is quickly attacked by hot dilute inorganic acid
E insoluble in alkalis	E soluble in alkalis

Source: PEREIRA Jr. *et* al. (2008)

2.4.3 Lignin

It consists of a complex polymer with an amorphous structure, aromatic and aliphatic components, which is associated with cellulose and hemicellulose during the formation of the plant cell wall and its purpose is to give it rigidity. Its concentration in the fibres influences the structure, properties, morphology, flexibility and rate of hydrolysis (CARDOSO, 2008; SILVEIRA, 2008).

Lignin is a complex macromolecule formed by the radical polymerisation of phenyl-propane units, which are three aromatic alcohols (Figure 5): p-coumaryl, p-coniferyl and p-sinapyl (MELO, 2010; CARDOSO, 2008).

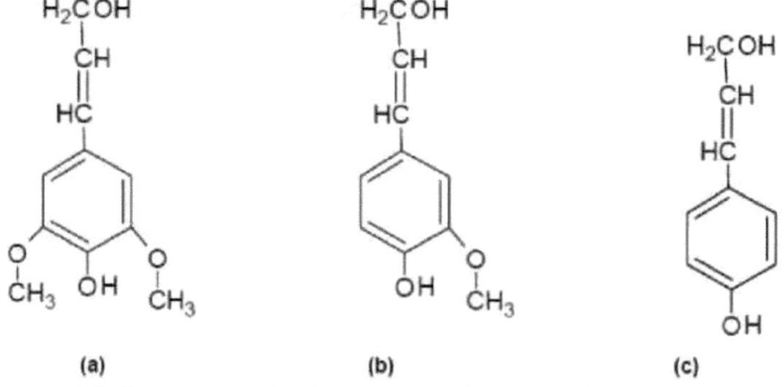

Figure 5 - Lignin precursor molecular structures: (a) p-sinapyl alcohol, (b) p-coniferyl alcohol and (c) p-coumaryl alcohol

The degradation of lignin is of great economic importance, making available substances of interest to industry, livestock farming and agriculture.

Lignin can also be used in the production of phenolic resins or can be gasified with oxygen, providing synthesis gas, which is essential in the production of methanol and can be used as an important compound in the production of a wide variety of chemical products (SCHUCHARDT & RIBEIRO, 2001).

2.5 PRE-TREATMENT OF LIGNOCELLULOSIC MATERIALS

Lignocellulosic materials are mainly composed of cellulose, hemicellulose and lignin in a complex structure that is resistant to degradation. The problems faced in the pre-treatment stage of converting biomass into fermentable sugars are justified by the low accessibility of cellulose due to its rigid association with lignin, as shown in Figure 6 (GUPTA *et* al, 2010). In other words, pre-treatment is a crucial and costly unit process in the conversion of lignocellulosic materials into biofuels.

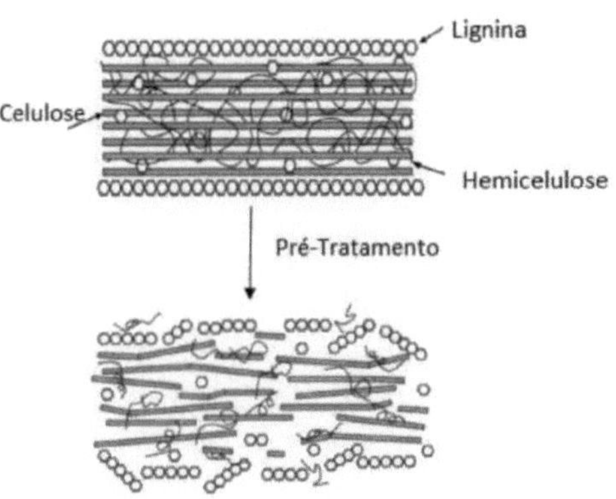

Figure 6-Structures of lignocellulosic materials before and after pre-treatment
Source: MOOD *et* al, 2013

A suitable pre-treatment procedure involves breaking the hydrogen bond in crystalline cellulose, decomposing the hemicelluloses and lignin, and finally increasing the porosity *and* surface area of the cellulose for the subsequent step (LI et al, 2010). There are various pretreatment methods including physical treatments (crushing and grinding, microwaves and extrusion), chemical pretreatments (alkaline, acid, organosolv, ozonolysis and ionic liquids), physicochemical (steam explosion, hydrothermal, fibre explosion with ammonia, wet oxidation and CO explosion$_2$) and biological treatments.

2.5.1 Acid pre-treatment

Acid pretreatments, specifically those using sulphuric acid as a catalyst, are the most commonly used for lignocellulosic materials in which polysaccharides, especially hemicelluloses, are hydrolysed into monosaccharides, facilitating the accessibility of the cellulose molecule.

This type of pre-treatment can be carried out using low acid concentrations and high temperatures or high acid concentrations and low temperatures (TAHERZADEH et al, 2008). Industrially, dilute acids are more attractive because low concentrations of fermentation inhibitors are generated during this stage of the process.

Pinheiro et al (2011) evaluated the thermochemical pre-treatment of sugar cane bagasse for the production of fermentable sugars and used the following conditions to obtain a hydrolysis efficiency of 57.40%: temperature of 120°C, sulphuric acid concentration of 0.5M and reaction time of 1 minute. Campos *et* al (2013) studied the acid pre-treatment of sugar cane with $H_2 SO_4$ at a concentration of 0.75% (v/v), at a temperature of 120°C for 45 minutes. Under these conditions, the most efficient process was obtained for solubilising the hemicellulose and preserving the structure.

2.3.2 Alkaline pre-treatment

The main functions of this type of pre-treatment are to remove lignin, acetyl groups and different uronic acids that can inhibit accessibility to cellulose for subsequent saccharification (LI et al, 2010). The solubilisation of hemicelluloses and cellulose in this method is lower than in acid and hydrothermal processes (CAVALHEIRO et al, 2008).

According to Mosier et al (2009), alkaline pre-treatment can be carried out under ambient conditions, with the caveat that this process will require more time than conventional pre-treatment when pressure and temperature are used.

Sodium hydroxide (NaOH), potassium hydroxide (KOH), calcium hydroxide (CaOH) and ammonia (NH_3) are the most commonly used reagents in this type of pre-treatment. Oliveira et al. (2010) studied alkaline pre-treatment using NaOH (1%) at 100°C for 60 min, resulting in 80.0% delignification of sugarcane straw. Nascimento (2011), working with sugar cane bagasse using 7% NaOH for 30 min at 120°C, obtained 95% delignification of the biomass.

2.3.3 Combination of acid and alkaline pre-treatments

Recently, combined pre-treatments have been considered a promising means of overcoming the challenge of increasing the efficiency of sugar production from the degradation of polysaccharides, since these combinations aim to expose the cellulose molecule to facilitate the next phase, which is hydrolysis. Various combinations have been carried out, for example, combining alkaline and ionic liquid pretreatments, acid and steam explosion, supercritical CO_2 and steam, organosolv and biological, acid and alkaline, among others, the latter being the most commonly used in current research.

Acid pre-treatment could delignify and increase the surface area of the cellulose fibre, however, a single stage pre-treatment with acetic acid, for example, requires a concentration of around 50%, based on the initial amount of dry material, so to overcome this challenge, a combined pre-treatment could be applied to partially, also remove some of the lignin in a single stage (MOOD *et* al, 2013).

Mendes et al. (2012) evaluated seven different pre-treatment strategies for *Brachiaria brizantha* cv. Marandu lignocellulosic biomass and the acid pre-treatment (H2SO4) followed by the basic pre-treatment (NaOH), both at 4%, showed the best results among those evaluated, resulting in a material with 92.36% cellulose, and its enzymatic hydrolysis converted 94.62% of the cellulose into glucose.

Filho (2014), when studying the production of cellulosic ethanol from fodder palm, carried out an acid pretreatment (H2SO4) followed by a basic pretreatment (NaOH), both at 0.5 per cent, and obtained a cellulose recovery of around 32.31 per cent and a percentage of 24.19 per cent delignification of the lignocellulosic material.

2.6   CATALYSED HYDROLYSIS

The catalytic hydrolysis stage takes place after pre-treatment of the lignocellulosic material and is where hemicellulose and cellulose are degraded to release sugars. This process can be carried out using different methods, including biological or chemical. Chemical hydrolysis usually uses sulphuric or hydrochloric acids in low or high concentrations. The main disadvantage of this method is the final cost of the product, as it requires the use of highly corrosion-resistant equipment due to the high concentration of acid. With regard to hydrolysis with dilute acids, it can be seen that high temperatures are required in this process, which will result in a higher concentration of furfural and also acetic acid that is generated from the hydrolysis of the acetyl groups of the hemicellulose. To a lesser extent, cellulose degradation can also occur, generating hydroxymethylfurfural (HMF) as a product. These by-products can inhibit the fermentation stage.

In biological hydrolysis, which is commonly carried out using cellulolytic enzymes, milder conditions are used compared to chemical hydrolysis. Temperatures close to 50°C are usually applied, which reduces the amount of unwanted by-products such as those mentioned above. This type of process has gained importance in the academic world due to the high rates of conversion into sugars that are obtained. However, the main disadvantage of this process is the use of cellulolytic enzymes, which currently have a high cost, and also the need for a long time to convert the cellulose.

As an alternative to the process of hydrolysis of the cellulose molecule, heteropolyacids, transition metal oxides, zeolites and mesoporous materials have been showing advantages, such as the ability to be reused, the low level of wear and tear on industrial equipment, the lower generation of effluents, as well as the fact that the liquor obtained does not require neutralisation.

Tong *et* al (2012) studied the application of a series of solid catalysts to the hydrolysis of cellulose into reducing sugars in aqueous media, including zeolite H, montmorillonite and acid-activated montmorillonite. When evaluating the activity of the catalysts, the authors found that montmorillonite treated with phosphoric (H_3PO_4), p-toluenesulphonic (PTSA) and sulphuric (H_2SO_4) acids showed the highest catalytic activity, with 78.5, 83.3 and 91.2%, respectively, being achieved in cellulose conversion.

Given these new perspectives for the hydrolysis of cellulose, catalytic systems using clay minerals have gained significant importance. This potential can be justified by the high selectivity of these materials associated with their probable reuse, contributing mainly to environmental issues. In addition, solid catalysts can be more advantageous because of their easy separation from the products, the possibility of reuse, as well as having important elements for their modification, such as high surface area, a large number of hydroxyl groups and high stability.

CHAPTER 3

MATERIALS AND METHODS

3.1   RAW MATERIAL

This study used sisal fibres from Fazenda Montevidéu in the municipality of Nova Floresta, PB, and the clay used in this work is palygorsquita (atapulgite), provided by the company Mineração Coimbra LTDA, extracted from a deposit located in the municipality of Guadalupe, PI.

3.2   PHYSICAL PRE-TREATMENT

The sisal fibres were ground in a Marconi MA048 knife mill, Figure 7, to 30 mesh. After the milling process, the fibres were dried and characterised.

Figure 7- Knife mill

3.3   PHYSICO-CHEMICAL CHARACTERISATION OF SISAL FIBRE

To characterise the fibres *in natura* and pre-treated, a methodology was used based on the procedures for lignocellulosic materials, document 236 EMBRAPA

(MORAIS et al., 2010). The materials were characterised in terms of moisture content, ash, extractives, lignin, holocellulose, alpha-cellulose and hemicellulose.

3.4 ACID ACTIVATION OF ATAPULGITE CLAY

The atapulgite clay samples were activated with hydrochloric acid according to the methodology of Rodrigues et al. (2006). Figure 8 shows the summarised acid activation flowchart for atapulgite clay.

Figure 8 - Flowchart of the acid activation methodology for atapulgite clay.

3.5 CLAY CHARACTERISATION

The clays in their natural and chemically treated forms were characterised by X-ray diffraction (XRD) and physical adsorption of N2 by BET.

3.5.1 X-ray diffraction

The X-ray curves of the samples were obtained using the powder method. The tests were carried out on a Shimadzu XRD 6000 diffractometer. A CuKa radiation source (λ =1.5406 Å) was used, obtained by 40kV current with a 30 mA filament. The measurements were carried out with a step of 0.02°, with a sweep of $5° \leq 2\Theta \leq 50°$. This analysis was carried out at the Ceramic Materials Synthesis Laboratory at the Federal University of Campina Grande.

3.5.2 Physical Adsorption of N2 by BET

The textural analysis of the clay was carried out using a Quantachrome Nova 3200e surface area and pore size analyser. This analysis was carried out at the Ceramic Materials Synthesis Laboratory of the Materials Engineering Academic Unit of the

Federal University of Campina Grande (UFCG).

3.6   ACID PRE-TREATMENT

The acid pre-treatment was based on the study carried out by Leão (2014) in which he assessed the potential for obtaining ethanol from sisal fibre. The pre-treatment was carried out in a MAITEC FORNOS model INTI stainless steel reactor with a pressure and temperature controller and a maximum capacity of 750 mL, as shown in Figure 9.

Figure 9 - Stainless steel reactor used for acid pre-treatment.

To carry out the acid pre-treatment, the reactor was fed a ratio of 1/10 dry mass of fibre/volume of aqueous sulphuric acid solution at a concentration of 3% at a temperature of 120°C for a period of 60 min. The material was then filtered and washed with distilled water until the pH equalled that of the wash water, after which the material was dried in an oven at 60°C for 24 hours.

3.7   ALKALINE PRE-TREATMENT

For the alkaline pre-treatment, sisal fibre pre-treated with sulphuric acid under the conditions described above was used. At this stage, the reactor was fed a ratio of 1/10 dry mass of pre-treated fibre/volume of aqueous sodium hydroxide solution at a concentration of 4% at a temperature of 120°C for 60 min. The material was then filtered and washed with distilled water until the pH was equal to that of the wash water, after which the material was dried in an oven at 60°C for 24 hours. The acid pre-treatment followed by the alkaline pre-treatment generated the material used for hydrolysis using atapulgite clay as a catalyst.

3.8   CATALYTIC HYDROLYSIS

The hydrolysis of sisal fibre was carried out in polytetrafluoroethylene autoclaves lined with stainless steel parts, with a maximum capacity of 25 mL. The methodology used in this process was adapted for the sisal fibre raw material, using the methodology employed in the hydrolysis of sugar cane bagasse using vermiculite clay as a catalyst (NUNES *et* al, 2014).

A 2-factor design[3] was used with three experiments at the centre point to check the influence of the input variables (process temperature (T), substrate mass (m) and time (t)) on the glucose concentration. In this process, the volume of distilled water was set at 20 ml and the mass of atapulgite clay activated with hydrochloric acid at 1.0 g. Table 3 shows the full factorial design matrix and Table 4 the coded and real levels of the input variables.

Table 3-Factorial planning matrix 2^3

Experiment	Temperature (T)	Substrate mass (m)	Time (t)
1	- 1	-1	-1
2	+ 1	-1	-1
3	- 1	+ 1	-1
4	+ 1	+ 1	-1
5	- 1	-1	+1
6	+ 1	-1	+1
7	- 1	+ 1	+1
8	+ 1	+ 1	+1
9	0	0	0
10	0	0	0
11	0	0	0

Table 4- Coded and actual levels of the independent variables for the planning

Variables	-1	0	+1
Temperature (°C)	160	180	200
Substrate mass(g)	0,5	1,0	1,5
Time (h)	2	4	6

3.9   DETERMINATION OF REDUCING SUGARS

The concentration of reducing sugars was determined following the methodology described by Miller (1959), with adaptations, which is based on the reduction of 3.5 dinitrosalicylic acid (DNS) to 3-amino-5-nitrosalicylic acid, simultaneously with the oxidation of the aldehyde group of the sugar to a carboxylic group. The sugars were quantified by constructing a calibration curve that correlates the absorbance with the sugar concentration, expressed as glucose.

CHAPTER 4

RESULTS AND DISCUSSION

4.1 CHARACTERISATION OF LIGNOCELLULOSIC MATERIAL

The sisal fibres used in this study were characterised before and after the pre-treatment stage (acid followed by alkali). The data on moisture content, extractives, ash, lignin, holocellulose, alpha-cellulose and hemicellulose present *in the fresh* samples is shown in Table 5, as well as data on sisal fibres found in the literature.

Table 5- Comparison between the percentage of parameters analysed in this study and others described in the literature

Parameters	Fresh fibre (%)[0]	Fresh fibre (%)[1]	Fresh fibre (%)[2]	Fresh fibre (%)[3]
Humidity	7,28 ±0,17	Not evaluated	5,59 ± 0,32	Not evaluated
Extractives	9,90 ± 0,21	9,27	7,22 ± 0,06	Not evaluated
Haemicellulose	21,56 ± 1,20	24,58	15,21 ±2,14	19,0
Ashes	1,35 ±0,04	1,95	2,4 ±0,17	1,0
Lignin	14,92 ± 0,48	6,97	11, 08 ±0,17	12,3
Holocellulose	70,37 ± 0,80	83,75	73,61 ± 1,92	84,5
Alpha-cellulose	48,82 ± 1,21	59,18	58,40 ± 0,94	65,5

[0] This work
[1] JÚNIOR (2006)
[2] LEÃO (2014)
[3] SALAZAR *et* al (2006)

Looking at the results shown in Table 5, it can be seen that the composition of fresh sisal fibre was similar to that found by Salazar et al (2006) and Júnior (2006) with regard to hemicellulose and ash content, but there was a discrepancy with regard to holocellulose content, which can be explained by factors such as the climate in which it was grown, the method and time of planting and harvesting, as well as the different methods used to determine it.

However, when comparing the holocellulose and lignin content of this study with the results obtained by Leão (2014), both of whom used the same method for determination and the fibre came from the same growing site, there is a greater similarity.

When it came to the alpha-cellulose content, the lowest content was found, approximately 49 per cent, when compared to the other studies cited. The extractive content was in line with that found by Salazar *et* al (2006) and Leão (2014).

When we analyse the three main components of the chemical characterisation of the fibre used in this work, cellulose, hemicellulose and lignin, we can see that although this material has a significant cellulose content, it needs to go through a pre-treatment stage to remove some of the lignin and break down the hemicelluloses, increasing the cellulose content.

A new characterisation of the sisal fibre was carried out after the pre-treatment stage. The results obtained are shown in Table 6.

Table 6- Chemical composition of pre-treated fibre

Parameters	Pre-treated fibre (%)
Humidity	2,28 ±0,13
Extractives	4,7 ± 1,08
Ashes	1,35 ±0,02
Lignin	5,83 ± 0,42
Holocellulose	85,76 ± 1,49
Alpha-cellulose	76,22 ±0,12
Haemicellulose	10,39 ±0,02

It is important to note that after this stage, the moisture, extractive, lignin and hemicellulose contents decreased, while the holocellulose and alpha-cellulose values increased considerably. These results confirm that the expected objective of acid pretreatment followed by base pretreatment was achieved. A removal of around 61% of lignin, a solubilisation of approximately 52% of hemicellulose and a concentration of alpha-cellulose of 36%. For a better visualisation of the alpha-cellulose, hemicellulose and lignin contents before and after pre-treatment, Figure 10 illustrates the results mentioned above.

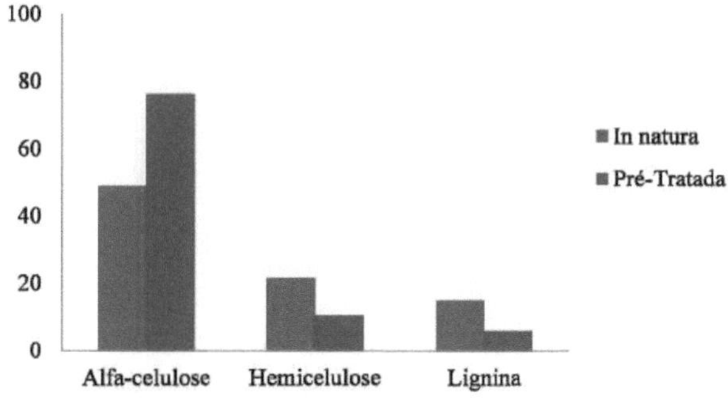

Figure 10 - Composition of sisal fibre

Table 7 shows a comparison of the cellulose content of sisal fibre with other pre-treated agro-industrial lignocellulosics used for bioconversion to ethanol evaluated in the literature.

Table 7 - Comparison between the percentage of cellulose found in this study and other pre-treated lignocellulosic materials described in the literature

Lignocellulosic waste	Cellulose (%)	Reference
Sisal fibre	76,22	This work
Sugar cane bagasse	58,64	Christofoletti (2010)
Banana tree pseudostem	55,80	Silva (2010)
Fodder Palm	45,38	Son (2014)
Sugarcane straw	51,90	Silva (2010)

As can be seen in Table 7, sisal fibre compared to other lignocellulosic materials described in the literature is promising for bioconversion into high added value products. When comparing the composition of the sisal used in this work with those described in the literature (Table 7), it can be seen that the amount of cellulose was higher than the other materials, reaching a difference of 30.84% from fodder palm.

4.2 CLAY CHARACTERISATION

4.2.1 X-ray diffraction

The X-ray diffraction curves of the natural and hydrochloric acid chemically treated atapulgite clays are shown in Figure 11.

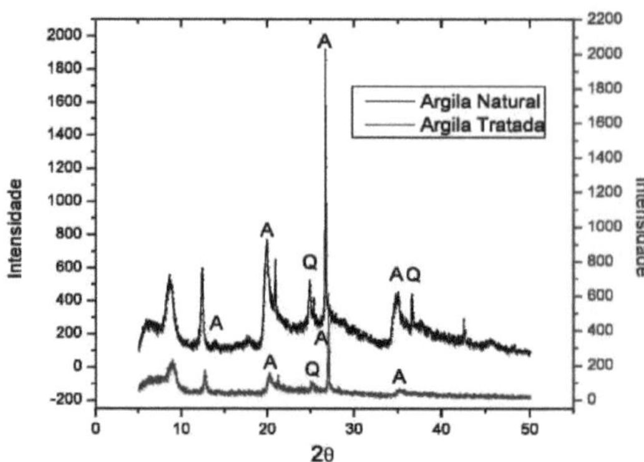

Figure 11 - X-ray diffraction curve for natural and treated atapulgite clays

The X-ray diffraction results show the peaks representing atapulgite clay and quartz in the natural and treated clays, where the main reflections of these materials were identified with the help of the crystallographic files No. 310783 and 82-1557, respectively.

The region referring to the highest intensity reflection of atapulgite $d_{(231)}$ is located at approximately 26.6°. Other reflections characteristic of this clay mineral, present in the diffractogram and of lower intensity, were also identified, such as d(2oo) at 13.92°, $d_{(121)}$ at 20.85° and $d_{(061)}$ at 34.95°, as well as the nearby region at 25.51° and 36.4°, which are characteristic of the presence of quartz, attributed to contamination, and are well documented in the literature by Xavier *et* al. (2012) and Sátiro (2013).

The results of the X-ray diffraction curves shown in Figure 11 show that after the acid treatment there was a decrease in the intensity of the characteristic peak of atapulgite when compared to the natural sample, which is in line with studies carried out by Santos (2013) and Oliveira (2011), who showed that adding variables such as mechanical agitation and consequently a longer contact time between the clay and the acid solution resulted in a decrease in the characteristic peak of atapulgite. In addition, it was found that after treatment there was a reduction in the intensity of the quartz peak by 25.51° and its elimination by 36.4°.

4.2.2 Physical adsorption   of N2 by BET

The textural characterisation of the sisal fibre was carried out using N2 adsorption and desorption. Gas adsorption and desorption isotherms make it possible to determine properties such as surface area, pore volume, morphology and pore size.

30

Figure 12 shows the adsorption isotherms of natural clay (a) and clay treated with hydrochloric acid (b).

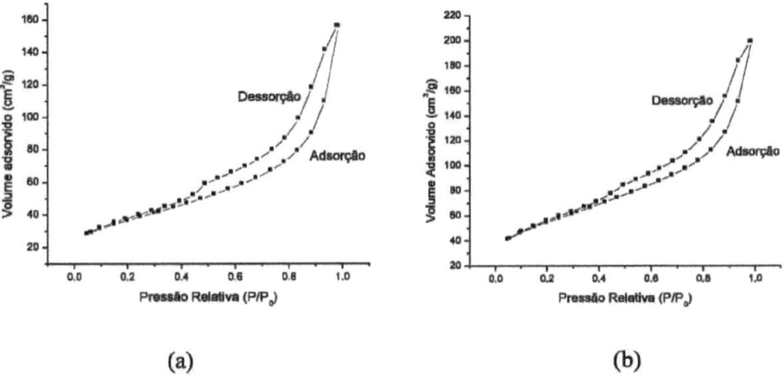

Figure 12- N2 adsorption isotherms of natural (a) and treated (b) clays

Figure 12 shows that, according to the IUPAC classification, the two isotherms obtained exhibit type IV behaviour, typical of mesoporous materials. This type of isotherm has well-defined levels which indicate capillary condensation; this level corresponds to the filling of all the pores with the adsorbate in the liquid state. According to the IUPAC classification, the two clay isotherms show the presence of type H3 hysteresis, typical of lamellar materials with slit-type pores.The surface area of the natural atapulgite sample showed a relatively high value of 127.5 $m^2 .g^{-1}$ as shown in Table 8. This value was higher when compared to other atapulgite clays found in the literature, a fact that can be attributed to the location where the clay was extracted.

Table 8- Comparison between the specific surface area values of atapulgite

Surface area ($m^2 .g$)$^{-1}$	Reference
127,5	This work
123	Zhang *et al.* (2010)
125,1	Boudriche *et al.* (2011)
113	Xavier et al. (2012)

Table 9 shows the surface area and pore volume results of the natural samples and those subjected to acid activation with HC1.

Table 9- Comparison between the surface area and pore volume of the natural and treated clays.

Sample	Surface area ($m^2 .g$	Pore volume ($cm^3 .g$

	$)^{-1}$	$)^{-1}$
Natural	127,5	0,2417
Treated	192,7	0,3084

Table 9 shows a significant increase (approximately 34%) in the specific surface area of atapulgite after chemical treatment and a 22% increase in pore volume. This phenomenon can be explained by the fact that the chemical treatment removed impurities from the surface of the clay, which corroborates the X-ray diffraction analysis discussed earlier.

4.3 CATALYTIC HYDROLYSIS OF SISAL FIBRE

The catalytic hydrolysis of the sisal fibre was carried out after the pre-treatment stage because, as seen above, the lignin and hemicellulose content is significant and could make it difficult to expose the cellulose to the hydrolytic agents. Table 10 shows the results of the concentrations of reducing sugars (RS) obtained in the hydrolysis for each experiment carried out in the experimental design, with emphasis on Experiment 8, where at a temperature of 200°C, a substrate mass of 1.5g and a time of 6 hours, the highest concentration of reducing sugars was obtained, 9684.4 mg.L^{-1} .

Table 10 - Concentration of reducing sugars obtained in experimental plan 2^3

Experiment	Temperature (°C)	Substrate mass (g)	Time (h)	AR(mg.L)$^{-1}$
1	- 1 (160)	- 1 (0,5)	-1(2)	162,35
2	+ 1(200)	-1(0,5)	-1(2)	563,16
3	- 1(160)	+ 1(1,5)	-1(2)	257,71
4	+ 1(200)	+ 1(1,5)	-1(2)	551,24
5	- 1(160)	-1(0,5)	+1(6)	1146,7
6	+ 1(200)	-1(0,5)	+1(6)	3933,0
7	- 1(160)	+ 1(1,5)	+1(6)	598,92
8	+ 1(200)	+ 1(1,5)	+1(6)	9684,4
9	0(180)	0(1,0)	0(4)	2249,3
10	0(180)	0(1,0)	0(4)	2130,1
11	0(180)	0(1,0)	0(4)	2383,4

The values obtained were added to the STATISTICA 12.0 programme. Equation 1 shows the response model for the concentration of reducing sugars and is given as a function of the coded values for the variables temperature (T), substrate mass (m) and hydrolysis time (t). The coefficients in bold are those that are significant at a 95% confidence level, as can be seen in the Pareto Diagram (Figure 13).

Cone. AR (mg.L^{-1}) = 2150.9 +

1570.8.T+660.9.m+1728.6.t+774.0.T.m+1397.2.T.t+640.0.m.t Equation (1)

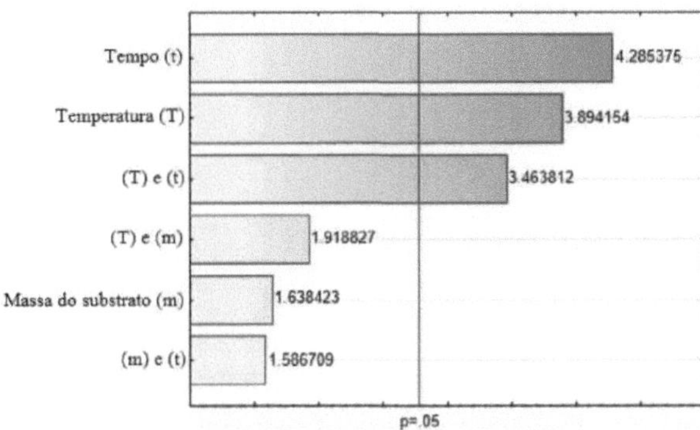

Figure 13 - Pareto diagram for hydrolysis planning

Table 11 shows the ANOVA of the RA concentration response.

Table 11 - ANOVA for the reduction sugar concentration response

Source of Variation	Sum of squares	Degrees of freedom	Quadratic mean	Fcalculated
Regression	70822561	6	11803760,17	9,068
Waste	5206503	4	1301625,75	
Total	76029064	10		

R^2 = 93.152%; F tabulated = 6.163

The data in Table 11 shows that the model is statistically significant at the 95 per cent confidence level, since the ratio of Fcalculated to Ftabulated is 1.47, and is above 1.0 (RODRIGUES EIEMMA, 2005).

As the response model under study is statistically significant, the response surface can be constructed (Figure 14). It should be noted that the model is not predictive as it lacks fit, but the surface can be analysed to look for trends in the influence of the input variables on the response.

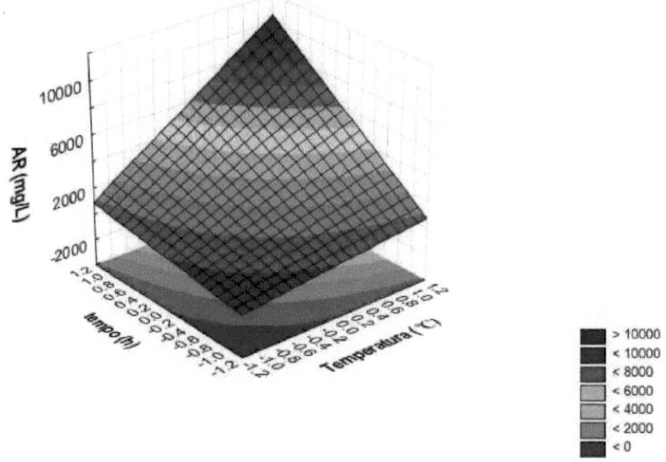

Figure 14 - Response surface: influence of the variables time (t) and temperature (T) on the concentration of reducing sugars, given the mass of substrate at level +1 (1.5g)

Analysing Figure 14, it can be seen that an increase in temperature and an increase in time leads to an increase in the concentration of reducing sugars, where operating at levels +1 (200°C) and +1 (6h) for these variables, respectively, results in concentrations of reducing sugars in the hydrolysed liquor above 9,684.0 mg.L^{-1} .

It should be noted that this value may be higher, since it was not possible to determine exactly what concentration of glucose was present in the liquor, due to the methodology used to quantify it, when compared with the results obtained by Nunes (2015), who working with the hydrolysis of sugarcane bagasse using vermiculite clay as a catalyst found a glucose concentration of around 2.112.0 mg.L^{-1} operating at 200°C for 2 hours and also higher than that found by Torres Neto (2015) who, studying the potential of fodder palm through acid hydrolysis for the production of second generation ethanol, managed to obtain a glucose concentration of 7,914.95 mg.L^{-1} at a temperature of 140°C, a sulphuric acid concentration of 3% and a palm/acid dry mass ratio of 1:9.

CHAPTER 5

CONCLUSIONS

From the results obtained from the study of the hydrolysis of sisal fibre using atapulgite clay as a catalyst, it can be concluded that:

- Sisal fibre is a raw material with great potential for producing fermentable sugars that can be used to produce second-generation ethanol, as it has a considerable cellulose content;

- The combination of acid and alkaline pre-treatment used on this lignocellulosic biomass was effective in breaking down the existing barrier in the cellulose fibres and increasing the cellulose content of the material, thus making it easier to expose the cellulose for degradation into sugars;

- The results of the techniques used to characterise the atapulgite from Guadalupe-PI showed that this clay mineral is highly crystalline, with the d(23i) reflection for atapulgite being more intense than that for quartz, as shown by the XRD analysis.

- The specific surface area analysis indicated that the clay mineral under study after chemical treatment showed an increase of approximately 34 per cent in specific surface area and 22 per cent in pore volume.

- The highest concentrations of reducing sugars found in catalytic hydrolysis occurred under the following conditions: temperature equal to 200°C, substrate mass 1.5g and reaction time of 6 hours.

5.1 SUGGESTIONS FOR FUTURE WORK

- Investigating new pre-treatment routes for sisal fibre;

- Use high-performance liquid chromatography to quantify the sugars obtained from the hydrolysis of sisal fibre;

- Chemically activate the atapulgite clay with other acids, such as phosphoric or sulphuric acids.

BIBLIOGRAPHICAL REFERENCES

AQUINO, D. F. **Sisal - Minimum Price Proposal 2012/2013: Internal Studies.** CONAB, 2012.

ARANTES, Valdeir; SADDLER, Jack N. Access to cellulose limits the efficiency of enzymatic hydrolysis: the role of amorphogenesis. **Biotechnol Biofuels,** [s.l.], v. 3, n. 1, p.4- 15, 2010.

BNDES. **Sugarcane-based ethanol for sustainable development.** Rio de Janeiro, p. 1-304.
<http ://www. sugarcanebioethanol.org/en/download/bioethanol .pdf>, 2008.

BOUDRICHE, Lilya et al. Effect of acid treatment on surface properties evolution of attapulgite clay: An application of inverse gas chromatography. **Colloids And Surfaces A: Physicochemical and Engineering Aspects,** [s.l.], v. 392, n. 1, p.45-54, dec. 2011.

BRIGATTI, M.; GALAN, E.; THENG, B. Chapter 2 Structures and Mineralogy of Clay Minerals. **Developments In Clay Science,** [s.l.], v. 1, n. 1, p. 19-86, 2006.

CAMPBELL, C. **Lecture on the sisal market.** In: Sisal Farming Seminar. Conceição do Coité. 2004.

CAMPOS, L.M.A.; PONTES, L.A.M.; CARVALHO, L.S.; CHEMMÉS, C.S.; GARRIDO, C.V.S.; SILVA, F.C.; SANTOS, J.M.G.M.; LEAL, S.C.S.; MERCANDELLI, S.S.SILVA, V.L. **Avaliação de Pré-tratamento do Bagaço de Cana-de-Açúcar com Acido Diluído para Produção de Etanol de Segunda Geração.** 53rd Brazilian Chemistry Congress. Rio de Janeiro/RJ, 2013.

CARDOSO, V. M. **Application of electron beam radiation as pre-treatment of sugar cane bagasse for enzymatic hydrolysis of cellulose.** Dissertation (Sciences in the Area of Nuclear Technology - Applications), Nuclear and Energy Research Institute - IPEN, São Paulo, 2008.

CARVALHEIRO F.; DUARTE, LC.; GIRIO FM. Hemicellulose biorefmeries: areview on biomass pretreatments . **Journal of Scientific and Industrial Research;67:849-64, 2008.**

CASTRO, A.M, JÚNIOR, N.P. Production, properties and application of cellulases in the hydrolysis of agro-industrial waste. **Quim. Nova,** Vol. 33, No. 1, 181-188, 2010.

CHEN, F.; LU, S.; WANG, E.; TSENG, K. Renewable energy in Taiwan. **Renewable and Sustainable Energy Reviews;** v. 14, p. 2029-2038,2010.

CHOVAU, S.; DEGRAUWE, D.; VAN DER BRUGGEN, B. Criticai analysis of techno-economic estimates for the production cost of lignocellulosic bio-ethanol. **Renewable and Sustainable Energy Reviews.** pp. 307-321. 2013.

CHRISTOFOLETTI, G. B. **Study of the effects of pre-treatment steps on the acid hydrolysis of sugarcane bagasse.** Master's dissertation, University of São Paulo, 2010.

CHUM, H.; FAAIJ, A.; MOREIRA,; BERNDES, J. G.; DHAMIJA, P.. Bioenergy. In **IPCC Special Report on Renewable Energy Sources and Climate Change Mitigation.**
Cambridge University Press, Cambridge, United Kingdom and New York, NY, USA. 2011

COELHO, A.C.V., SANTOS, P.S.,SANTOS,H.S. Special clays: what they are, characterisation and properties. **Química Nova.** v. 70, n.l, p.146-152,2007.

Cunha, T. S. A **Atividade Sisaleira e a sua Interferência Sobre o Bioma da Caatinga no Município de Valente - BA.** Serrinha: UNEB, 2010. Monograph (conclusion of degree course in Geography). Department of Education. State University of Bahia.2010.

DUTRA *et* al. Evaluation of the potential of clays from Rio Grande do Norte-Brazil. **Cerâmica Industrial,** v.ll, n.2, p. 42-46,2006.

EPE - Energy Research Company. **National Energy Balance.** Available at: https://ben.epe.gov.br/.

FERREIRA, LL. **Flexibility in the use of diesel or biodiesel, an approach using real options theory.** Master's dissertation. Rio de JaneiroRJ: Getulio Vargas Foundation; 2007.

FILHO, P.F.S., **Fodder palm (Opuntia fícus indica and Nopalea cochelillifera as raw material for the production of cellulosic ethanol and cellulotic enzymes.** Master's dissertation. Federal University of Rio Grande do Norte, Natal, 2014.

GEREMIAS, M. L.. **Characterisation of clays from the Paraná basin, in the south of Santa Catarina, for use in the manufacture of ceramic tiles.** 2003. 225 f. Thesis (Doctorate) - Polytechnic School of the University of São Paulo, São Paulo, 2003.

GOMEZ, E. O., SOUZA, R. T. G., ROCHA, G.J.M., ALMEIDA, E., CORTEZ, L. A.B. Evaluation of lignocellulosic residues for EG ethanol production. **Revista Analytica,** v.48, p. 180, 2010.

HUBER, G. W.; IBORRA, S.; CORMA, A. Synthesis of Transportation Fuels from Biomass: Chemistry, Catalysts and Engineering. **Chem. Rev.** v. 106, n. 9, p. 4044-4098, 2006.

IBORRA,C.V. et al. Characterisation of northem Patagonian betoiútes for pharmaceutical uses. **Applied Clay Science,** v. 31, p. 272-281,2006.

IEA - International Energy Agency. **World Energy Outlook.** IEA, Paris. 2013

KUMAR, P., BARRET, D.M.,DELWICHE; M. L. J. E., STROEVE, P. Methods for pretreatment of lignocellulosic biomass for efficient hydrolysis and biofuel production. **Ind. Engineering Chemical Research,** v. 48, p. 3713-3729,2009.

JÚNIOR, J.D.M. **Sisal fibres: Study of properties and chemical modifications for application in phenolic matrix composites.** Doctoral thesis. University of São Paulo. 272 p. 2006.

LEÃO, Douglas Alexandre Saraiva. **Potential for obtaining ethanol from sisal fibre.** 2014. 93 f. Thesis (Doctorate) - Process Engineering Course, Federal University of Campina Grande, Campina Grande, 2014.

LEITE, R.C.C.; LEAL, M.R.L. V; CORTEZ, L.A.B.; GRIFFIN, W.M.;SCANDIFFIO,M.I. G. Can Brazil replace 5% of the 2025 gasoline world demand with ethanol? **Energy Applied,** 34 (5),pp. 655-661,2009.

LI, X.; KIM, TH.; Nghiem, NP. Bioethanol production from com stover using aqueous ammonia pretreatment and two-phase simultaneous saccharification and fermentation (TPSSF). **Bioresource Technology**.p. 5910-5916, 2010

LI, Y.; MAI; Y. W. & YE, L. Sisal fibre and its composites: a review of recent developments. **Composites Sei. Technol.,** 6, p.2037, 2000.

LI, JINBAO; ZANG, XIANGRONG; ZANG, MEIYUN; XIU, HUIJUAN; HE, HANG. Ultrasonic enhance acid hydrolysis selectivity of cellulose with HCl-FeC13 as catalyst. **Carbohydrate Polymers.** Vol.117, p. 917-922, 2014.

LIMA, A. O. S.; RODRIGUES, A. L. Saccharification of cellulosic waste with recombinant bacteria as a strategy for reducing the greenhouse effect. **Revista de ciências ambientais,** v. 1, n. 2, p. 5-18, 2007.

LUZ, Adão Benvindo da; BALTAR, Carlos Adolpho Magalhães; OLIVERIA, Cristiano Honório de. Mineralogical and technological characterisation of atapulgites from Piauí. **Mineral Inputs for Oil Well Drilling.** Rio de Janeiro: Cetem, 2003. p.

82-99.

MELO, L. F. L.; **Chemical substances in effluents from the extraction of cellulose nanocrystals.** Course Conclusion Paper (Bachelor's Degree in Chemistry), Federal University of Ceará, Fortaleza, 2010.

MENDES, T. D.; PACHECO, T. F.; CARVALHO, F.B.P.; NAKAI, D. K.; RODRIGUES, D. S.; MACHADO, C. M. M., AYRES M. **Evaluation of different pre-treatments for the deconstruction of lignocellulosic biomass.** Proceedings of the XIX Brazilian Congress of Chemical Engineering. Búzios - RJ, 2012.

MILLER, G.L. Use of dinitrosalicylic acid reagent for determination of reducing sugar.
Analytical Chemistry. 31, 426-428. 1959.

MOOD, S. H.; GOLFESHAN, A. H.; TABATABAEI M.; JOUZANI, G. S.; NAJAFI, G.H; GHOLAMI, M.;ARDJMAND, M. Lignocellulosic biomass to bioethanol, a comprehensive review with a focus on pretreatment. **Renewable and Sustainable Energy Reviews.** V. 27, p.77-93, 2013.

MORAIS, S.A.L.; NASCIMENTO, E.A.; MELO, D.C. Chemical analysis of Pinusoocarpa wood PARTI - Quantification of macromolecular components and volatile extractives. **Revista Árvore,** v. 29, n. 3, p. 461-470, 2005.

MORAIS, J.P.S.; ROSA, J. M; MARCONCINI, M. F. **Procedures for lignocellulosic analysis,** Campina Grande: Embrapa Algodão, Documentos, 236, 54 p, 2010.

MOSIER, N.; WYMAN, C; DALE, B.; ELANDER, R.; LEE, Y.Y.; HOLTZAPLLE, M.; LADISCH,M. Features of promising technologies for pretreatment of lignocellulosic biomass. **Bioresource Technology,** v. 96, p. 673-686,2009.

MURRAY, H.H. Traditional and new applications for kaolin, smectite, and palygorskite: a general overview. **Applied Clay Science,** vol.17, p. 207-221,2000.

NASCIMENTO, V.M. **Alkaline pretreatment of sugarcane bagasse for ethanol production and obtaining xylooligomers.** Master's dissertation. Federal University of São Carlos. São Paulo. 126 p, 2011.

NAVES, I.M. **Sisal - 2012/2013 harvest: commercialisation - proposed actions.** Report. 18p. 2012

NUNES, B. R. P.; CONRADO, L.S.; MORAIS, C.R. **Study of the pre-treatment and hydrolysis of sugarcane bagasse using acid-activated vermiculite as a catalyst.** XX Brazilian Congress of Chemical Engineering. Florianópolis-SC. Annals of the congress, 2014.

NUNES, B. R. P. **Evaluation of the use of vermiculite clay as a catalyst in the hydrolysis process of sugarcane bagasse.** 123 f. Thesis (Doctorate) - Chemical Engineering Course, Federal University of Campina Grande, Campina Grande, 2015.

OLIVEIRA, R. S, **Treatment and characterisation of atapulgite for use in face masks and for reinforcement in composites with PVA.** Dissertation in Metallurgical and Materials Engineering, COPPE, Federal University of Rio de Janeiro, 2011.

OLIVEIRA, F.M.V., **Evaluation of different pre-treatments and alkaline delignification in the saccharification of cellulose from sugarcane straw.** Master's dissertation. Lorena School of Engineering, University of São Paulo. 98 p. 2010

PEREIRA Jr" N.; COUTO, M.A.P.G.; SANTA, L.M.M. Biomass of lignocellulosic composition for fuel ethanol production and the context of biorefmery. In **Series on Biotechnology,** Ed. Amiga Digital UFRJ, Rio de Janeiro, v.2, 45 p, 2008.

PINHEIRO, F.G.C; COSTA, A.G.; MORAIS, J.P.S.; SANTOS, A.B.; SANTAELLA, S.T.;
LEITÃO, R.C. **Thermochemical pretreatment of sugarcane bagasse for the production of fermentable sugars.** II International Symposium on Agricultural and Agroindustrial Waste Management, 2011.

RODRIGUES, J.A.R. **Do engenho à biorrefmaria: A sugar mill as an industrial enterprise for the generation of biochemical products and biofuels.** Institute of Chemistry, University of Campinas, Campinas - SP, Brazil, 2010.

RODRIGUES, L.D. **Sugarcane as a raw material for biofuel production: environmental impacts and agro-ecological zoning as a tool**

for mitigation. Final coursework in environmental analysis. Federal University of Juiz de Fora. 59 p. 2010.

RODRIGUES, M. I. and IEMMA, A. F. (2005), **Planning experiments and optimising processes.** Casa do Pão Editora, 325p.

RODRIGUES, M. G; PEREIRA, K. R. O; VALENZUELA DÍAZ, F. R. Obtaining and characterising chemically activated clay materials for use in catalysis. **Ceramics,** n.52, p.260-263, 2006.

SÁ, Lívian R. Vasconcelos de; CAMMAROTA, Magali C.; FERREIRA-LEITÃO, Viridiana S.. Hydrogen production by anaerobic fermentation - general aspects and possibility of using brazilian agro-industrial wastes. **Química Nova,** [s.l.], v. 37, n. 5, p.857-867, 2014.

SALAZAR, V.L.P.; LEAO, A.L. Biodegradation of coconut and sisal fibres applied in the automotive industry. **Energia Agrícola.21 (2): 99-133. 2006**

SANTOS, M. S.F. **Analysis of the action of polygorschite in the treatment of effluents contaminated by lead.** PhD thesis. Federal University of Campina Grande, Campina Grande, 2013.

SÁTIRO, M.B., **Biodiesel obtained from soya and cottonseed oil using atapulgite clay.** Master's dissertation. Federal University of Campina Grande, Campina Grande, 2013.

SCHUCHARDT, U.; RIBEIRO, M. L. The petrochemical industry in the next century: how to replace oil as a raw material. **Química Nova.** v. 24, n. 2, p. 247-251, 2001.

SILVA, L. C. A., **Obtaining and characterising nanocomposites based on polyhydroxyalkanoate/atapulgite.** Dissertation in Materials Science and Engineering.
Federal University of Sergipe, São Cristovão, 2010.

SILVA, V. F. V. N. **Studies of pre-treatment and enzymatic saccharification of agro-industrial residues as steps in the process of obtaining cellulosic ethanol.** Master's dissertation, University of São Paulo, Lorena, Brazil. 2010.

SILVA, A. S.; SILVA, F.L.H.; CARVALHO, W.N.C; PEREIRA, K.R.O.; LIMA, E.E. Cellulose hydrolysis using NiO-MCM-41 and MoO3-MCM-41 mesostructured catalysts.
Química Nova. Vol.35, p. 683-688. 2012.

SILVEIRA, M. S. **Utilisation of green coconut shells for briquette production in Salvador - BA.** Dissertation (Master's in Environmental Management and Technologies in the Production Process). Federal University of Bahia, Salvador, 2008.

SOARES, D. S., **Evaluation of atapulgite clay for potential use as a pharmaceutical excipient in solid forms.** Master's dissertation. Federal University of Rio Grande do Norte, São Natal, 2013.

TAHERZADEH, M.J. & KARIMI, K. Pretreatment of Lignocellulosic Wastes to Improve Ethanol and Biogas Production: A Review. **International Journal of Molecular Sciences,** 9, pp.1621-1651. 2008.

THE ECONOMIST. **Brazil Takes Off.** November 14, 2009.

TOLMASQUIM, M.T. Perspectives and planning of the energy sector in Brazil. **Estudos Avançados;26(74):247-60, 2012.**

TONG, D. S. Catalytic hydrolysis of cellulose to reducing sugar over acid-activated montmorillonite catalysts. **Applied Clay Science** p. 147-153. V. 74,2012.

TORRES NETO, A.B. **Production of bioethanol from the acid and enzymatic hydrolysis of fodder palm.** 93 f. Thesis (Doctorate) - Chemical Engineering Course, Federal University of Campina Grande, Campina Grande, 2015.

XAVIER, K. C. M.; SILVA FILHO, E C.; SANTOS, M. S. F.; SANTOS, M. R. M. C.; LUZ, A. B.. Mineralogical, morphological and surface characterisation of atapulgite from Guadalupe- PI. **HOLOS,** Year 28, Vol5. P 60-70, 2012.

YANG, Q.; LUE, A; Zhang, L. Reinforcement of ramie fibres on regenerated cellulose films. **Composities Science and Techonology** 70: 2319-2324. 2010.

ZHANG, J.,WANG, Q.,CHEN, H.,WANG, A.; XRF and nitrogen adsorption studies of acid- activated palygorskite. **Clay Minerals,** v. 45, n. 2, p. 145-156, 2010.

Buy your books fast and straightforward online - at one of world's fastest growing online book stores! Environmentally sound due to Print-on-Demand technologies.

Buy your books online at
www.morebooks.shop

Kaufen Sie Ihre Bücher schnell und unkompliziert online – auf einer der am schnellsten wachsenden Buchhandelsplattformen weltweit! Dank Print-On-Demand umwelt- und ressourcenschonend produzi ert.

Bücher schneller online kaufen
www.morebooks.shop

Printed by Books on Demand GmbH, Norderstedt / Germany